故園畫憶

庚寅中秋

韩启德题

故园画忆系列
Memory of the Old Home in Sketches

石家庄漫记
Shijiazhuang Album

宋端　绘画 撰文
Sketches & Notes by Song Duan

學苑出版社
Academy Press

图书在版编目（CIP）数据

石家庄漫记 / 宋端绘画、撰文. -- 北京：学苑出版社，2013.6
（故园画忆系列）
ISBN 978-7-5077-4291-6

Ⅰ. ①石… Ⅱ. ①宋… Ⅲ. ①建筑画－钢笔画－作品集－中国－现代 Ⅳ. ①TU-881.2

中国版本图书馆CIP数据核字(2013)第113951号

出 版 人：孟 白
出版发行：学苑出版社
社　　址：北京市丰台区南方庄2号院1号楼
邮政编码：100079
网　　址：www.book001.com
电子信箱：xueyuan@public.bta.net.cn
销售电话：010-67675512、67678944、67601101（邮购）
经　　销：全国新华书店
印 刷 厂：北京信彩瑞禾印刷厂
开本尺寸：889×1194 1/24
印　　张：6
字　　数：120千字
图　　幅：119幅
版　　次：2013年6月北京第1版
印　　次：2013年6月北京第1次印刷
定　　价：38.00元

目　录

乡土漫记

Contents

Country Album

Red Album

Folk-custom Album

序

前段时间宋端和我联系说正在画一本书，是一个叫“故园画忆”的系列，希望我到时能给他写序，我欣然接受。近日他将已画好的两本画儿拿给我看，当如此大量的关于石家庄的建筑速写呈现在我眼前时，我不禁由衷地赞叹。在看画儿的过程中，宋端向我详细地介绍着每一幅画儿的信息，这些建筑的始建年代、结构、装饰，以及现在的保护情况，他都如数家珍。他不但让我了解到很多石家庄的古迹和民俗知识，同时也为更多读者提供了认识这些传统文化的一扇窗。所以我说宋端的这本《故园画忆——石家庄漫记》所呈现给读者的绝不仅仅是一本钢笔速写集。

宋端平时也从事环境艺术设计的教学工作，在这之前我看过他画的皖南徽派建筑的钢笔速写，画风清新利落。今次再看到新作，发现他的钢笔速写在风格上更加娴熟，且能根据不同题材、不同场景，进行不同风格的呈现。他并没有在书中固守一个风格，也没有放置一张老画，所有作品均为从去年五月至今所画。在这之前我看过很多钢笔速写集，大多是讲解技法等知识的书，像这样以传播文化为视角的钢笔速写集并不多。宋端是个认真的人，对专业一丝不苟。为此他利用工作之余几乎走遍了石家庄所辖的所有区县，同时查阅了很多史籍和文献资料。

在看这些钢笔速写的过程中，我发现这些画儿的角度也不尽相同。宋端对我说他不想让画面千篇一律，因此每一幅速写在构图和角度的选择上，不一定会选择正面或最漂亮的角度画，均尽量选择最有利于呈现该古迹面貌的角度进行写生，尽量让画面完整。在宋端这些钢笔速写中，我仿佛真的看到了这些古迹，也对石家庄的历史古迹与民俗文化有了新的认识。

宋端多次和我说他不希望钢笔速写仅仅是当做一种表现技法而存在，他希望通过钢笔速写传递一种文化。而通过他的这本钢笔速写，我想我已看到了文化。

2013年5月

Preface

A while ago Song Duan contacted me and said he was creating a book, which was called "Memory of the Old Home in Sketches", and hoped I could write the preface, a request I was pleased to accept. He had drawn two sketchbooks of paintings and brought them to me recently. When I saw so many sketches of the Shijiazhuang buildings, I felt a mixture of both pride and delight. While looking at these sketches, Song Duan told me detailed information about each building's age, structure, decoration, and conservation status. Not only did he introduce me to a lot of the monuments and folk knowledge of Shijiazhuang, but he also provides a window of this traditional culture for more readers. So I consider the "Memory of the Old Home in Sketches-- Shijiazhuang Album " illustrated by Song Duan, is not just a sketchbook but an informative text.

Song Duan also teaches environmental art and design. Prior to this work, he has done sketches of the buildings in the south of Anhui, with a fresh and neat style. In this book, I can see that he has become even more proficient in his work, and he is capable of varying his styles according to different themes and scenes. All of his works were painted over the course of the last year. I had seen a lot of the pen sketches before, most of them were explain the technique used, but this book is about the dissemination of culture with in addition to the sketches. Song Duan is a serious person and a meticulous professional. In order to finish this book, he took advantage of his spare time and has been to almost all the districts and counties under the jurisdiction of Shijiazhuang, additionally, he referred to a lot of history books and local literature.

Looking at these pen sketches, I found that they are not at the same angle. Song Duan told me he did not want to make the pictures to seem monotonous, so he tried to sketch the buildings from various angles and compositions in order to make the picture complete. These sketches not only illustrate the monuments but highlight the folk culture of Shijiazhuang.

Song Duan had told me many times that he does not want the pen sketches to only be seen as a manifestation of techniques, rather he hopes to convey the culture of the region through these illustrations. Throughout this book, I believe I have seen that culture.

Guan Shangqing
May 2013

前　言

这本书用图画的方式对石家庄的古建筑、石家庄的民居、石家庄的红色岁月，以及石家庄特有的一些民俗文化，进行“漫记式”记录。其实从2011年春天我即开始了本书的筹备，工作之余我用了近一年的时间走访了石家庄数十处古迹和村落，参观很多本地的民俗活动。并将这些掠影用钢笔速写形式的进行了展现。

石家庄一直以来都给人年轻城市的印象。其实石家庄历史悠久，辖区内有国家级重点文物保护单位25处，省级文物保护单位81处，文物单位保有量在全国名列前茅。同样，石家庄的地域民俗文化也独具特色，我亦选择了一部分进行了展现。

其实早在本书的创作之初，学苑出版社的周鼎老师就建议我可以把所有画作进行一个分类，按类别将书列为几部分，这给了我很大启迪。经多次调整后，我将本书分为古建漫记、民居漫记、红色漫记和民俗漫记四个部分。

古建漫记是本书的第一部分，也是篇幅最多的部分，石家庄各级文保单位众多，其中包含各种遗址、古建筑等，而古建筑无疑是其中最耀眼的，部分古建筑还是国内孤品。长期以来石家庄多给人缺少历史的新兴城市印象，这显然是不对的。我在这一部分的编排中，将全国文保单位的古建筑、桥梁等置于前列，之后为省级文保单位和其他古建筑。并将这些古迹按照其所在的区域进行排列，使之尽量有序。本书的第二部分是民居漫记，石家庄的民居从材料上大体分为两部分，一是石砌建筑，二是砖木建筑。石家庄位于太行山东侧，而河北一带从元朝开始即是京畿之地，故石家庄的民居风格既有山西建筑的影子，也有北京四合院的风貌。石家庄的传统民居保护较好的，是太行山区的井陉县的民居。第三部分我将描绘的重点安排为红色文化，石家庄是革命圣地西柏坡的所在地，故本书将这些红色文化单列。而本书的第四部分，为民俗部分。主要从曲艺类、歌舞、社火、庙会、饮食、手工艺等多个视角展示石家庄的特色民俗文化。借助钢笔速写这个载体和这次出版的机会，我也对石家庄的古迹和传统文化有了更深的认识。

最后我要感谢学苑出版社周鼎老师的信任，是他为我提供了写这本书的机会。我还要感谢我的同学——郑州大学教师王安安的举荐。没有他们就不会有本书的出版。同时，还要感谢我的老师——中国著名画家关尚卿先生能在百忙之中为本书作序。

宋端

2013年5月

Author's Foreword

This book is an album of Shijiazhuang, its ancient buildings, dwellings, its red years, and characteristic folk culture . From the spring of 2011, I began the preparation of this book , and I spent nearly a year of spare time, visiting Shijiazhuang dozens of historical sites and villages, as well as observing many local folk activities . I provide a glimpse of these activities in the form of the pen sketches in this book.

Shijiazhuang has been giving the impression of the young city . In fact , Shijiazhuang has a long history within the jurisdiction of the national key cultural relics protection units 25 , 81 of the provincial cultural relics protection units of cultural relics in population among the best in the country . Meanwhile, Shijiazhuang area also has a unique folk culture, which I also attempt to show.

In fact, as early as the beginning of the creation of this book, the Academy Press Zhou Ding teacher advised me to divide the book into categories and assign each illustration to a category. After several adjustments , I divided the book into four parts that reflect the ancient architecture, dwellings, red years and folk culture .

Ancient architecture album is the first part of the book, it is also the largest section. Shijiazhuang has a large amount of cultural protection units at all levels, which includes a variety of sites. Ancient buildings are undoubtedly one of the most dazzling part of the landscape of Shijiazhuang. Shijiazhuang does not always showcase its rich history. As a result of this the national cultural protection unit of ancient buildings, bridges, etc. has made an effort to conserve and showcase these buildings. Monuments are arranged according to the their location, in order to make the book as orderly as possible. The second part of the book takes a look at the dwellings in Shijiazhuang. The material can be divided into two parts, one is stone buildings, and the other is brick-wood buildings. Shijiazhuang is located in the east of the Taihang Mountains, from the Yuan Dynasty , Shijiazhuang's residential style is reminiscent of the Shanxi and Beijing Siheyuan architectural style. The Jingxing Shijiazhuang better protection of traditional houses of the Taihang Mountains, residential part of Jingxing majority. In the third part I will depict the Red Culture. Shijiazhuang is the location of the sacred place of revolution xibaipo, Shijiazhuang City, the liberation of the People's Liberation Army a big city , so the book make the red culture a single row . The fourth part of the book is the folk culture section, and showcases the dance, community, temples, food, crafts and other characteristics of folk culture in Shijiazhuang . My travels and sketches have also enabled me to gain a greater understanding of Shijiazhuang monuments and traditional culture .

Finally, I would like to thank Zhou Ding teacher of the Academy Press, he provided me the opportunity to write this book . I would also like to thank my classmates — the teachers of the University of Zhengzhou, Wang Anan, for her nomination and recommendation. This book would not have been published without their help. At the same time, I would also like to thank my teacher, the famous Chinese painter Guan Shangqing, who took the time to write the preface for his book despite his busy schedule.

Song Duan

May 2013

岁月漫记

Time Album

毗卢寺（新华区）

始建于唐朝天宝年间（742～756年）。该寺主殿毗卢殿建在高一米的月台上，前后有抱厦，平面呈十字形，因其形制特殊，俗称五花八角殿。

Pilu Temple (Xinhua District)

It was built in the Tianbao reign of the Tang Dynasty (742~756). The main shrine of Pilu Temple was built on one meter high platform and the plane of it was criss-cross.It was commonly called streaky octagonal temple because of its special shapes.

永安桥（新华区）

建于明万历年间（1573～1619年），大桥为单孔石拱桥，桥为南北走向，全部用青石条砌筑而成，桥面长20米，宽4.5米，拱长11米。

Yongan Bridge(Xinhua District)

Built in the Ming dynasty Wanli years (1573-1619), the bridge is a single-hole stone arch bridge, the bridge is North-South, all with stone masonry, and bridge deck is 20 meters long, 4.5 meters wide, Arch length of 11 meters.

石家庄大石桥（新华区）

建于1907年，全长150米，高七米，宽十米。桥身为石灰岩砌成，是石家庄历史的见证。

Shijiazhuang Dashi Bridge (Xinhua District)

It was founded in 1907, and the total length was 150 meters, 7 meters high, 10 meters wide. the bridge was the witness of Shijiazhuang's history, which was built with limestone.

〔正太饭店·正门（新华区）〕

始建于1907年，是石家庄现存唯一的法式小洋楼，是当时正太铁路通车的服务性配套设施，在石家庄的历史上具有非常重要的地位。

Zhengtai Hotel · Main Gate(Xinhua District)

It was founded in 1907 and the only existing French small building in Shijiazhuang, which had been the service facility of Zhengtai railroad traffic at that time, and had a very important position in the history of Shijiazhuang.

正太饭店·院内（新华区）

是一座欧式风格建筑，也是当时石家庄最豪华的饭店，曾接待过孙中山、蒋介石、冯玉祥等人。

Zhengtai Hotel · Courtyard(Xinhua District)

It was a European style building, had received Sun Zhongshan, Jiang Jieshi, Feng Yuxiang et al, and was the most luxurious hotel in Shijiazhuang at that time .

懋华亭（新华区）

1935年6月为纪念正太铁路收归国有而建，亭高9米，钢筋混凝土结构。“懋华”二字取自为收回路权做出贡献的王懋功与朱华两人的名字。

Maohua Pavilion(Xinhua District)

It was built to commemorated the nationalisation right of Zheng Tai railway, the pavilion is 9 meters high, reinforced concrete structure. “Mao Hua” come from the name of Wang Maogong and Zhu Hua who made a contribution to the right of railway.

崇宁真君庙牌楼（桥西区）

又称关帝庙牌楼，建于1541年，牌楼有四根柱子组成，木额枋前后有八个题字，前为“浩然正气”，背面为“山西夫子”。

Chongning Zhenjun Temple Archway (Qiaoxi District)

It also known as the Guan Yu Temple's Gate, built in 1541, consisted of four pillars, before and after the wooden architrave of the archway had been inscribed with "Hao Ran Zheng Qin" and " Shanxi Fuzi".

民生路（桥东区）

形成于1915年，取名公兴存街，1946年更名为民生街，是石家庄建城初期一条主要街道，是汇聚各行各业的商业中心。

Minsheng Road (Qiaodong District)

Minsheng Road formed in 1915, named Gongxingcun street, and changed its name to Minsheng Road in 1946 .It had been the main street in the period of found the Shijiazhuang city earlier, and was the commercial center which cluster of all professions and trades.

米家楼（桥东区）

位于煤市路，砖混结构，原建于民国时期，为米姓商人所建。宽大拱圈门洞居中，房屋布局均衡，圆拱窗砌造精细，彰显建筑的稳重与大气。

Mijia Building (Qiaodong District)

It lies in the Meishi road, brick mix structure, built by the merchant whose family name is "Mi" in the period of the Republic of China. The large arch door opening was placed in the middle, housing balance layout, arched windows built fine, and showed the stable and magnificent temperament of the building.

民生客栈（桥东区）

位于原民生路南一巷，砖木结构。该房汇集青砖、拱圈、简瓦、檐口等北方传统建筑元素，呈典型民国时期建筑风格。

Ming Sheng Inn (Qiaodong District)

It lies in the Nanyixiang, Minsheng Road, brick-wood structure. The Inn has a collection of blue brick, arch, simple tiles, cornices and other northern traditional architectural elements, was typical of the period of the Republic of China Architectural style.

南寨清凉阁（井陉矿区）

又名南寨阁，建于1540年。建筑坐北朝南，占地800平米，房屋25间，有青石砌成三米高的台基。

South Village Qingliang Pavilion (Jingxing Mining District)

Also known as the South Village Pavilion, built in 1540, facing south, covered an area of 800 square meters, 25 rooms, and had a 3 meters high platform with blue brick.

段家楼群（井陉矿区）

建于1913年，是段祺瑞（北洋政府陆军总长兼代总理）及其家族在井陉矿区正丰煤矿兴建的一座具有西洋建筑风格的段家楼群。

Duanjia Building Group (Jingxing Mining District)

Duanjia building group was built in 1913 , founded by Duan Qirui (Army Chief of the northern government and Prime Minister) and his family in the Jingxing Mining District in Zhengfeng coal mine, with western architectural style.

伏羲台遗址（新乐市）

是新石器时代遗址，相传为伏羲氏族发展壮大的地方，伏羲台是伏羲时代留下的遗物和遗迹，现存建筑为后世纪念伏羲而建。

Fuxitai Ruins (Xinle City)

It is the ruins of the Neolithic age, according to legend, this is the residence of the Fuxi's Clan. Fuxitai was the relics and ruins in Fuxi era, and the existing building was built in memory of Fuxi.

土门关（鹿泉市）

即古代的井陉口，是山西、陕西从古驿道通往华北的必经之路，具有十分重要的军事价值，为历代兵家必争之地，是楚汉井径之战的主战场。

Tumen Pass (Luquan City)

It was Jingxingkou in the ancient times and the only way which must be passed from Shanxi and Shaanxi to North China, and had very important military value, which Ancient battle ground and the main battlefield of Chu-Han caliper battle.

铁行会馆（鹿泉市）

又称东会馆，始建于1724年，是一座二进院落，原是专门从事铁器营销的商务中心。该会馆是河北省规模最大，保存最好的商业会馆类古建筑。

Tiexing Guild Hall (Luquan City)

It was also known as the East guild hall, a two courtyard building, built in 1724, and originally specialized in the marketing of iron goods. The guild hall is the largest and best preserved ancient building business hall in Hebei Province.

正定城·长乐门（正定县）

始建于北周，1449年扩建为周长二十四里，高三丈二尺的土城，1571年将土城改为砖城。长乐门为南门，是正定目前保存最好的一个城门。

Zhengding City · Changle Gate (Zhengding County)

It was built in the Northern Zhou Dynasty (557-581) , the expansion of it made the soil city's the circumference to twenty-four lis, three Zhangs two feet in 1449. And the soil city was changed to brick city in 1571. Changle gate is the south gate of it and the best preserved city gate in Zhengding at present.

正定城·镇远门（正定县）

正定城有四城门，东曰迎旭，南曰长乐，西曰镇远，北曰永安，并均附有月城和瓮城。镇远门城楼、瓮城等已无，仅存城门。

Zhengding City · Zhenyuan Gate(Zhengding County)

Zhengding city has four gates, East gate is Yingxu gate, South gate is Changle gate, West gate is Zhenyuan gate, North gate is Yongan gate, and has Yuecheng and Wongcheng. Zhenyuan gate tower and Weng Cheng have been ruined except for the gate.

临济寺·澄灵塔（正定县）

临济寺是佛教临济宗（禅宗五家之一）的祖庭。澄灵塔建于867年，为义玄大宗师衣钵塔，塔通高30.7米，是一座砖砌八角九级塔。

Linji Temple · Chengling Pagoda(Zhengding County)

The Linji temple is a place of Buddhism of Linjizong(one of the five Zen)'s origin. Chengling Temple was built in 867 years, as Yi Xuan grandmaster mantle tower, 30.7 meters high, a brick octagonal nine floors pagoda.

开元寺·三门楼（正定县）

开元寺始建于540年，三门楼位于开元寺前部，始建于692年，现仅存石柱、栏额等。

Kaiyuan Temple · Sanmen Building (Zhengding County)

Kaiyuan Temple was built in 540 years. Sanmen building lies in the front of the temple, and was built in 692 years, but there are only remaining pillars and Lanes at present.

开元寺·须弥塔（正定县）

又称雁塔，始建于636年。塔身建在高约1.5米的正方形砖砌台基上，塔平面为正方形，密檐9级，塔高39.5米。

Kaiyuan Temple · Xumi Pagoda (Zhengding County)

Also known as the Yan pagoda, built in 636 years. The pagoda was built on the square brick platform which about 1.5 meters high, and the plane of it was square. It was 9 floors, 39.5 meters high.

开元寺·钟楼（正定县）

为砖木结构的二层楼阁式建筑，建筑面积135平方米，通高14米。其大木结构、柱网、斗拱都展示了唐代建筑艺术风格，是中国现存唯一的唐代钟楼实例。

Kaiyuan Temple · Bell Tower (Zhengding County)

It was two floors pavilion-style building with brick and wood structure, construction area of 135 square meters, 14 meters high. The large wooden structure, column grid and dougong showed artistic style of Tang Dynasty architecture, what's more, it is the only existing examples of Tang Dynasty tower in China at present.

隆兴寺·天王殿（正定县）

隆兴寺始建于隋朝，中国十大名寺之一。天王殿，单檐歇山式，七檀中柱式建筑，中有圆拱形大门，门上部横嵌着康熙皇帝亲书的“敕建隆兴寺”金字匾额。

Longxing Temple · King Hall(Zhengding County)

Longxing Temple was built in the Sui Dynasty, one of China's ten famous temples. The King Hall is a building which has single eaves jehiel mountain type and seven Tan in column, with arched door in the middle. And the gold plaque on the top of door was written by Emperor Kangxi, "Chi Jian Longxing Temple" .

隆兴寺·摩尼殿（正定县）

是该寺主要建筑之一，坐落在中轴线前部，始建于1052年。大殿结构属抬梁式木结构，平面呈十字形，体现了宋代建筑的特点。

Longxing Temple · Mani Temple(Zhengding County)

The temple is one of the main buildings in Longxing Temple, located in the front of the axis, founded in 1052. The temple is Tailiang style and wooden structure, the plane was cross-shaped, embodies the characteristics of the Song Dynasty architecture.

隆兴寺·戒坛（正定县）

平面呈方形的多檐亭台式建筑，是举行授戒仪式的场所，建于清乾隆年间(1736～1795年)。坛建在高约二米的两层须弥基座上，四面中间各有九级台阶。

Longxing Temple · Jietan (Zhengding County)

Jietan was a square of the multiple eaves pavilion-type building, held the Shoujie ceremony, built in the Qianlong period (1736 ~ 1795). The Jietan was built on the two floors Sumeru base which about 2 meters high, the middle of four sides had 9 steps.

隆兴寺·悲氏阁（正定县）

悲氏阁与转轮藏阁相对峙立于主体建筑大悲阁前，均为北宋时期的二层楼阁。该阁外观为单檐歇山顶。

Longxing Temple·Beishi Pavilion (Zhengding County)

Beishi Pavilion and Zhuanlunzang Pavilion stand face to face and are both standed in front of the main building -Dabei Pavilion, were the two floors pavilions in the Northern Song dynasty. The appearance of the pavilion was a single eave and Xieshan roof.

隆兴寺·大悲阁（正定县）

又名佛香阁、天宁观音阁，始建于宋开宝年间（968～976年)。大悲阁内矗立着一尊高大的宋代铜铸大菩萨，通高22米有余，是中国现存古代铜像中最高的一座。

Longxing Temple·Dabei Pavilion (Zhengding County)

Also known as Foxiang Pavilion or Tianning Guanyin Pavilion, built in the Kaibao reign of the Song Dynasty (968 ~ 976). There is a great bronze statue of Bodhisattva in Dabei Pavilion, which was more than 22 meters high, and it is the tallest one of China's existing ancient bronze statue at present.

隆兴寺·毗卢殿（正定县）

原位于正定城内崇因寺内，殿内有一尊明朝铸的铜质毗卢佛像，被誉为中国古代最精美的铜铸毗卢佛。

Longxing Temple · Pilu Hall (Zhengding County)

It lied in the Chongyin Temple in the ancient city of Zhengding, and there was a bronze statue of Pilu Buddha in it. The bronze statue of Pilu Buddha is known as the most exquisite bronze statue in ancient China.

广惠寺华塔（正定县）

始建于唐代，现存华塔为金大定年间（1161～1196年）重修后的重要遗存，后历代均有修葺。该塔是现存华塔中造型最奇特、装饰最华美的一座。

Guanghui Temple · Hua Pagoda (Zhengding County)

It was built in the Tang Dynasty, rebuilt in the Dading reign of the Jin Dynasty(1161~ 1196), both ancient repaired, and it is the important remains at present. The pagoda is in the shape of the existing tower strange, decorate the most beautiful one.

天宁寺凌霄塔（正定县）

凌霄塔建于唐懿宗威通年间(860～874年)，是一座砖木结构的九层楼阁式塔，平面呈八角形，高41米。

Tianning Temple · Lingxiao Pagoda(Zhengding County)

The Lingxiao pagoda was built in Weitong reign of the Tang Dynasty (860 ~ 874), the nine floors pavilion-style building with brick and wood structure, plane was octagonal, 41 meters high.

正定文庙（正定县）

有我国现存最早的文庙大成殿。1933年建筑教育家梁思成先生考察时，将其定为唐末五代遗物。面宽五间，进深三间，面积650平方米。

The Confucius Temple in Zhengding(Zhengding County)

It has the earliest existing Dacheng Hall of Confucious'temple in China. In 1933, when the architectural educator Mr. Liang Sicheng researching it, and predicated that it was the relic of the end of Tang Dynasties. The Confucius Temple was five rooms wide, three rooms deep, and an area of 650 square meters.

赵云庙（正定县）

现存赵云庙是在清道光年间（1821～1850年）基础上重新修缮的，占地12亩。总建筑面积1500平方米。

Zhaoyun Temple (Zhengding County)

Zhaoyun Temple was restored on the base which built in Daoguang reign of the Qing Dynasty(1821 ~ 1850), covered an area of 12 mous,and the total construction area of 1500 square meters.

深泽真武庙（深泽县）

位于深泽县北极台上，北极台始建于1562年，约2160平方米，台高14米，台上建真武庙，供奉北极紫微大帝。

Zhenwu Temple in Shenze (Shenze County)

It is located on Beiji station in Shenze county , the station was built in 1562, about 2160 square meters, 14 meters high. And Zhenwu Temple was built on the station, worshiped Beiji Ziwei Emperor.

柏林寺塔（赵县）

塔坐落于柏林禅寺内，柏林禅寺始建于汉建安年间(196～220年)，是禅宗史上的一座重要祖庭。柏林寺塔为纪念唐代著名高僧从谂真际禅师而建造。

Bolin Temple Pagoda (Zhao County)

The pagoda was located in the Bolin Temple. The Bolin Temple was built in the Jianan reign of Han Dynasty (196 ~ 220), and also an important Zuting in Zen history. Bolin Temple pagoda was built to commemorate the hierarch Cong Shenzhenji, a famous Tang Dynasty Buddhist monk.

赵州陀罗尼经幢（赵县）

建于1038年，经幢全部用花岗岩石雕琢叠砌而成，外观造型酷似塔，所以当地俗称“石塔”，是我国现存最高大的石刻经幢。

Zhaozhou Dharani Jingchuang (Zhao County)

Founded in 1038, built with granite stone. Its appearance like a tower, so the locals commonly known the Jingchuang as the “Stone Tower”. It is the tallest stone Jingzhuang in China.

大观圣作之碑（赵县）

立于1108年，碑文为宋徽宗赵佶所撰，故称“圣作”。碑由青石雕成，高5.6米，碑额为当时的权相蔡京所题。

Daguan Shengzuo Stele (Zhao County)

Founded in 1108, the inscription was written by Zhao Ji, the Emperor Song Huizong, so called "Shengzuo". The stele made of blue stone, 5.6 meters high, and the Beie was written by the prime minister, Cai Jing.

宋村遗址（赵县）

宋村遗址面积约9000平方米，有厚达五米商周文化层。遗址内现存1274年建的西林寺塔，高约14米，四层，平面六边形。

Song Village Ruins (Zhao County)

Song Village ruins area of about 9000 square meters, with 5 meters thick layer of the culture of Shang and Zhou. Xilin Temple pagoda, which built in 1274, was in the ruins.The pagoda was 14 meters high, 4 floors, the plane of it is hexagonal.

赵州桥（赵县）

又名安济桥，建于隋大业年间（605～618年），由李春设计和建造，是世界上现存最早、保存最完善的古代敞肩石拱桥，有“天下第一桥”的美誉。

Zhaozhou Bridge (Zhao County)

Also known as Anji Bridge, built in the Daye reign of Sui Dynasty (605 ~ 618). It was designed and built by Li Chun, and was the world's earliest, best preserved ancient stone arch and open shoulder bridge, also be called “The World Bridge” .

永通桥（赵县）

又名小石桥，建于765年，是一座带腹拱单孔并列券敞肩式石桥，全长32米，宽6.3米，跨度26米。

Yongtong Bridge (Zhao County)

Also known as Xiaoshi Bridge, built in 765 years, a arch of single hole, the spacious shoulder - stone bridge, a total length of 32 meters, 6.3 meters wide, 26 meters span.

栾城文庙（栾城县）

始建于1299年，明、清均有修缮，为一层木结构，单檐歇山式古建筑，长20米，宽9.3米，大成殿前有精致的蟠龙丹陛踏步。

The Confucius Temple of Luancheng (Luancheng County)

Founded in 1299, had repaired in Ming and Qing Dynasty, a layer of wood structure, Single eaves, Xieshan style of ancient buildings, 20 meters long, 9.3 meters wide. And there were exquisite steps in the front of the Dacheng Palace of The Confucius Temple.

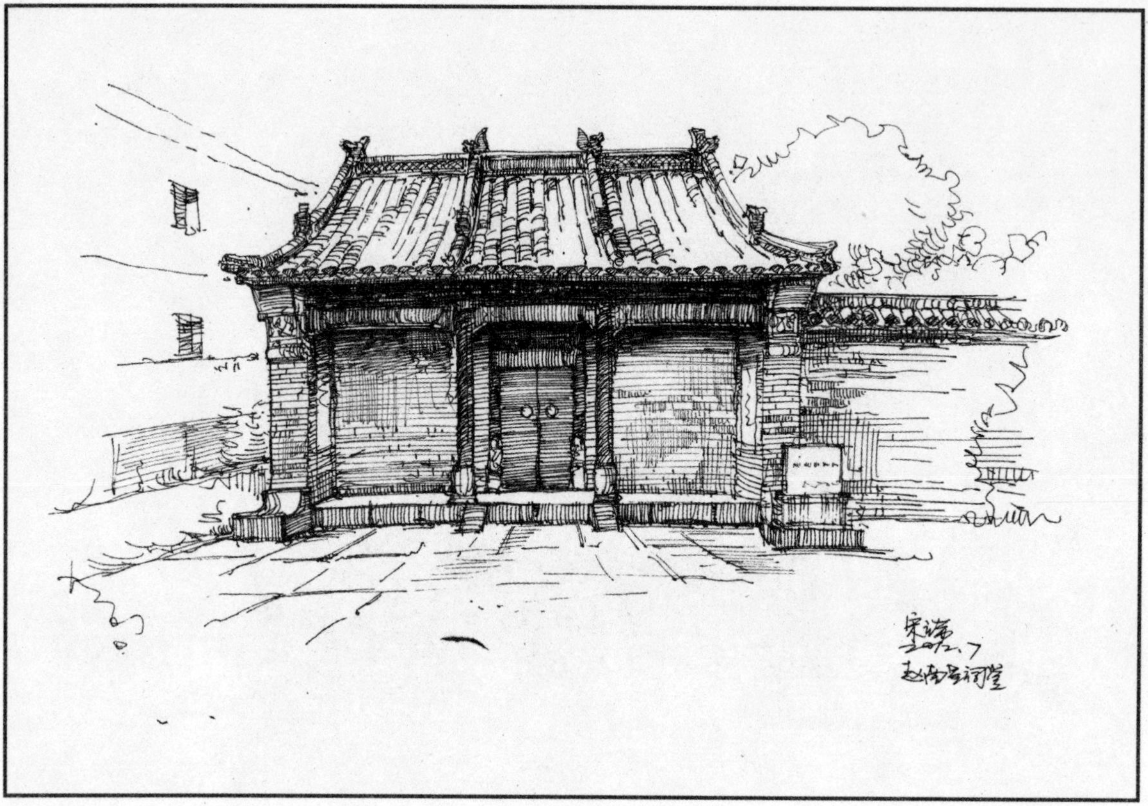

赵南星祠堂（高邑县）

始建于明代，清嘉庆年间（1796～1820年）重修。原称为赵氏宗祠，近代改称赵南星祠堂。

Zhao Nanxing Ancestral Hall (Gaoyi County)

Built in the Ming Dynasty, and rebuilt in the Jiaqing reign of Qing Dynasty (1796 ~ 1820) . Originally called Zhao ancestral hall, and was renamed the Zhao Nanxing ancestral hall in modern times.

开化寺塔（元氏县）

始建于金代，位于原开化寺内，现在寺院已无存，独存此塔。塔通高23米，塔身平面呈八角形，九级，为实心密檐砖塔。

Kaihua Temple Pagoda (Yuanshi County)

Built in the Jin Dynasty, lies in the former Kaihua Temple. Except for the pagoda, none of them is preserved now. The pagoda is 23 meters high, 9 floors, the plane of the pagoda was octagonal, a solid brick pagoda of Miyan style .

治平寺石塔（赞皇县）

又名嘉应寺石塔，始建于749年，塔平面呈八角形，均为青石雕砌，高约12.5米，塔身分四层，各层均出檐。

Zhiping Temple Pagoda (Zanhuang County)

Also known as the Jiaying Temple Pagoda, built in 749 years, octagonal tower with bluestone ornament, about 12.5 meters high, four layers, each layer had eaves.

井陉旧城·城门（井陉县）

位于井陉县天长镇，始建于明洪武年间（1368～1398年），初夯土城墙，隆庆六年(1572年)外包砖石，并在南门处设瓮城。

Jingxing Old City · City Gate(Jingxing County)

Jingxing County is located in the Tianchang Town, built in the Hongwu reign of Ming Dynasty(1368 ~ 1398), had been rammed earth walls in the early time. In Longqing six years (in 1572) , the wall was covered with brick and stone, and built a Wongcheng in the South Gate.

井陉旧城·东关（井陉县）

即天长镇东关，是井陉旧城保存较好的区域，现有的城门洞依然顺畅通行。城门洞及城墙下半部砌石，上半部砌砖。

Jingxing Old City · Dong Guan (Jingxing County)

Also known as the Dong Guan of Tianchang Town, a better preservation area, and the gate hole of Dong Guan still can be passed, and second half of the gate hole and the wall was stone, the upper half was brick.

井陉古驿道（井陉县）

是冀晋间太行山腹地的一条交通要道，兵家必争之地，素有“冀晋通衢”之称。

Ancient Post Road in Jingxing (Jingxing County)

It is a main traffic communications line in Taihang Mountain hinterland between Heibei and Shanxi province, a road of great military improtance and commonly known as “Ji Jin Tong Qu”.

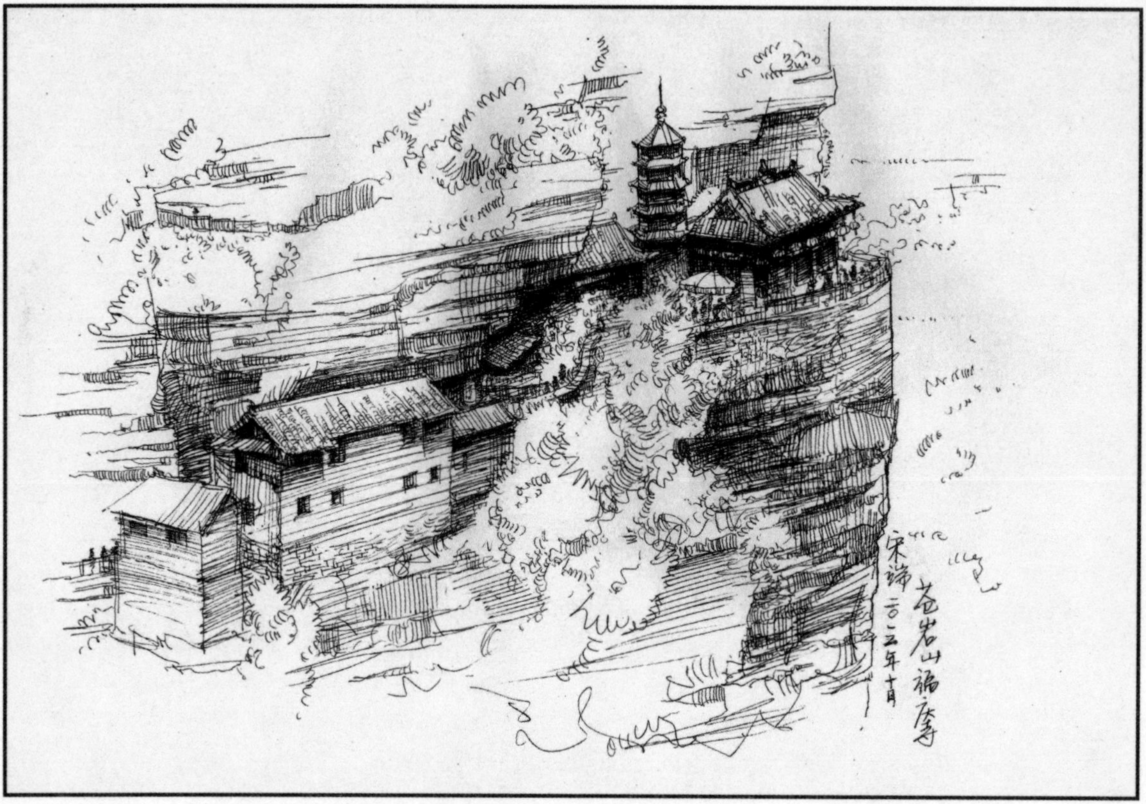

福庆寺（井陉县）

原名兴善寺，始建于隋朝，1014年宋真宗赐 “福庆寺”。现存建筑大都是清代遗物。

Fuqing Temple (Jingxing County)

Formerly known as the Xingshan temple, built in the Sui Dynasty, and Emperor Song Zhenzong called it "Fuqing Temple"in 1014. The existing building is mostly the relics built in Qing Dynasty.

福庆寺·桥楼殿（井陉县）

桥楼殿为我国三大悬空寺之一。石桥为隋代建筑，略早于赵州安济桥。桥上的楼殿为唐代建筑。

Fuqing Temple · Bridge House (Jingxing County)

Bridge house, one of three big Hanging Temple in China.The stone bridge was built in the Sui Dynasty, slightly earlier than Zhaozhou bridge. The house on the bridge was built in the Tang Dynasty.

福庆寺·天王殿（井陉县）

又称小桥楼殿，桥与殿同时建造于金代大定年间（1162～1189年），为歇山黄绿琉璃瓦顶的重檐楼阁式大木建筑。面阔三间，进深二间。

Fuqing Temple · King Hall (Jingxing County)

Also known as Small Bridge House, the bridge and the hall both were built in the Dading reign of the Jin Dynasty (1162~1189).It was a large wooden building with Chongyan Xieshan type and Yellow-green glazed tile roof.Three rooms long, two rooms wide.

井陉城隍庙（井陉县）

始建于明代，1843年重修。中轴线上分布着山门、戏楼、大殿及东西配殿，占地面积2500平方米。

Chenghuang Temple of Jingxing(Jingxing County)

Built in the Ming Dynasty, rebuilt in 1843.There were the gate, opera tower, main hall and East-West side hall on the Central Axis. Chenghuang temple covers an area of 2500 square meters.

天长镇观音阁（井陉县）

是明清建筑，为砖木结构，位于井陉旧城天长镇城门之上。观音阁主殿上悬“慈航普度”和“佑显灵威”金字牌匾。

Guanyin Pavilion in Tianchang Town (Jingxing County)

It was the Ming and Qing Dynasties building, a brick-wood structure, located on the gate of the Tianchang town in old Jingxing City. There were two gold plaques, “Ce Hang Pu Du” and “You Xian Ling Wei”,hanging on the main hall of Guanyin Pavilion.

井陉文庙（井陉县）

始建于元代，由棂星门、大成殿、万世宗师牌楼、宗圣祠等组成，是一处规模较为宏大的古建筑群。现存大成殿及部分偏殿。

The Confucius Temple in Jingxing(Jingxing County)

It was built in the Yuan Dynasty, composed of Jiuxing Gate, Dacheng Hall and All master's Pailou, a grand scale of ancient building, and now only existing Dacheng Hall and side halls.

皆山书院（井陉县）

建于1777年，是一座坐北朝南的三进院落。1903年改书院为县立高等小学堂，是井陉县开办的第一公办所学校。

Jieshan Academy(Jingxing County)

Founded in 1777, a building that facing south and has three small yards. The academy was changed into a perfectural upper primary school in 1903, and the first public school in Jingxing county.

龙窝寺石窟（井陉县）

位于龙窝寺内，现寺已无存，仅存崖壁上的石窟。石窟主要存有宋、明两代石刻佛教造像及题刻。

Longwo Temple Grottoes (Jingxing County)

Located in Longwo Temple, the temple has no deposit, only has the cliff grottoes. The Grottoes are mainly graved the Buddhist statues and Inscription of Song and Ming Dynasty.

通济桥（井陉县）

始建于1083年，跨径10米，宽6米，桥身为三孔石桥，通体青石砌筑，中间一孔拱跨较大，另外桥上有精美浮雕的青石板护栏。

Tongji Bridge (Jingxing County)

Founded in 1083, 10 meters long, 6 meters wide, a three-hole stone bridge with bluestone. The span of arch of the middle hole was the largest of the three holes. And there was a exquisite bluestone fence on the Tongji Bridge.

万寿寺塔林·之一（平山县）

占地六万多平方米，原有高僧墓塔23座，现存13座。

Wanshou Temple Pagodas · 1(Pingshan County)

Wanshou Temple Pagodas covering an area of 60000 square meters, the original had 23 Mutas and the existing had 13 Mutas.

万寿寺塔林·之二（平山县）

墓塔上均有铭文，年代有1297年、1307年、1310年等。墓塔均为单层，塔身高挑，各面略有雕饰。

Wanshou Temple Pagodas · 2(Pingshan County)

All of the Muta have inscriptions, written in 1297, 1307, 1310 and so on. The Muta was monolayer and tall, and every side had slightly carving.

平山文庙（平山县）

始建于741年，1103年迁建于此。原建筑总面积约1.5万平米。大成殿面阔五间，进深三间，梁架保留原木形态，为元代建筑风格。

The Confucius Temple in Pingshan (Pingshan County)

Founded in 741 years, 1103 was moved to this site. The total construction covers an area of 15000 square meters. The Dacheng Hall was five rooms wide, three rooms deep,and the beam mount was keep as stocks, the Yuan Dynasty architectural style.

幽居寺塔（灵寿县）

始建于557年，塔身建于简朴的方形台座上，七层，高23米，是北朝砖塔的典型遗存。

Youju Temple Pagoda (Lingshou County)

Founded in 557 years, the Pagoda was built on a simple and unsophisticated square steps, seven layers, 23 meters high, a typical Northern Dynasty pagoda relics.

灵寿石牌坊（灵寿县）

是典型的明代石牌坊建筑，建于1641年，该牌坊平面呈“口”字形，通体用“灵寿绿”大理石仿木透雕而成。

The Stone Memorial Archway of Lingshou (Lingshou County)

It is a typical stone memorial archway of the Ming Dynasty, built in 1641, the plane was buccal glyph, carved with the “Lingshou green”marble.

行唐琉璃殿（行唐县）

又名琉璃庙，建于1623年，是供奉观世音菩萨的宗教建筑。是一座单檐歇山式建筑，高约七米，是国内现存的惟一一座古代琉璃庙宇。

Liuli Hall of Xingtang(Xingtang County)

Also known as the Liuli Temple, built in 1623, is the religious building which worship Guanyin Bodhisattva, a Danyan Xieshan type architecture, about 7 meters high, is the only existing ancient Liuli hall in China.

封崇寺（行唐县）

始建于556年，现仅存天王殿和大殿，大殿坐北朝南，面积169.5平方米，为单檐琉璃瓦歇山式建筑。

Feng Chong Temple (Xingtang County)

Founded in 556 years, and now the only remaining King Hall and Main Hall .The Main Hall faces south, covers an area of 169.5 square meters, with single eaves and belongs to Xieshan type building.

行唐文庙（行唐县）

始建于1625年，现仅存大成殿。大成殿建在一米高的砖台上，殿高九米，面积300平方米，殿前有正方形月台。

The Confucian Temple of Xingtang (Xingtang County)

The Confucian Temple dates back to 1625, and there was the only remaining Dacheng Hall now. Dacheng Hall built on one meter high platform, 9 meters high, covering an area of 300 square meters, and there was a square platform in the front of the hall.

乡土漫记

County Album

常家院（桥东区）

原建于清末民初，位于原民生路北一巷。该庭院布局规整，为典型的四合院，建筑为砖木结构，砖雕、木雕尤为精致。该院为常姓家族所建。

The Chang's Courtyard (Qiaodong District)

Originally built in the late Qing Dynasty, was located in Beiyixiang, the former Minsheng Road. The courtyard's layout was regular, a typical Siheyuan with brick-wood structure, the brick carving and wood carving was very delicate. The courtyard was built for the Chang Shi's family.

申后高氏民居（鹿泉市）

建于1925年，位于申后村，是一个典型的四合院式民居。装饰精美，以砖雕、石雕、木雕为代表的建筑装饰做工精致，堪称石家庄一绝。

Gao Shi's Residence in Shen Hou Village(Luquan City)

Founded in 1925, is located in Shen Hou Village, a typical Siheyuan dwelling. The building was decorated with brick carving, stone carving and wood carving , it was a exquisite workmanship and called "the special building" in Shijiazhuang.

王士珍旧居·正门（正定县）

王士珍，正定县人，曾任中华民国陆军总长、北洋政府国务总理等职。旧居建成于1920年左右，坐北朝南，为民国时期北方典型的四合院。

Wang Shizhen's Former Residence · Main Gate (Zhengding County)

Wang Shizhen, Zhengding County, the former army chief of the Republic of China , Prime Minister of the Northern Government and other staff. The former residence was built in 1920, facing south, the typical Siheyuan during the period of Republic of China north .

王士珍旧居·院落（正定县）

整个院落分为中东西三路，房屋200余间，现仅存中路。中路为二进四合院，前院正房五间，后院正房七间，房基为青石砌筑。

Wang Shizhen's Former Residence · Courtyard (Zhengding County)

The courtyard is divided in three ways, more than 200 houses, and there is the only remaining middle way now. The middle way was a courtyard of the two, the front yard had five rooms, the backyard had seven rooms.Its base was built with bluestone.

王家大院·院门（井陉县）

位于井陉县天长镇，始建于明朝末年，在民国初年对宅院在进行扩建，使其成为一组大型建筑群，现保存完好。

Wangjia Courtyard · Main Gate (Jingxing County)

Jingxing county is located in Tianchang town, was built in the end of the Ming Dynasty. The Wangjia Courtyard was enlarged in the early years of the Republic of China, are well preserved till now.

王家大院·二门（井陉县）

大院占地6880平方米，房屋162间。该图展示了王家大院二门的样貌。

Wangjia Courtyard · Second gate (Jingxing County)

The Wangjia Courtyard covers an area of 6880 square meters, 162 rooms. The sketch shows the Second gate of Wangjia Courtyard.

王家大院·正房（井陉县）

建筑为石阶高台，青砖青瓦，门窗棂格，玲珑剔透，流露着明清之风。

Wangjia Courtyard · Principal room (Jingxing County)

Principal room was built on the stone base, with bluestone and grey tile. The door and windows of the principal room looks very exquisite and stable, the style during Ming and Qing dynasties.

天长镇街景（井陉县）

即井陉旧城关，始建于汉代，镇中古迹众多，其青砖青瓦的砖木结构民居也别具特色。

Tianchang Town Street (Jingxing County)

It was the old entrance to the county, built in the Han Dynasty, and there was a large number of antiquities in the town, especially the brick-wood folk houses had distinctive features which built with blue bricks and tiles.

于家村·村口（井陉县）

又称于家石头村，始建于明朝，是著名政治家、民族英雄于谦的后裔居所。于家村地处山区，故用于建筑上的主要材料都是山上的石头。

Yujia Village · The Entrance of the Village (Jingxing County)

The Yujia Village also known as the Yujia stone village, built in the Ming Dynasty. The descendants of Yu Qian lived in this village, who was a famous politician and national hero. Yujia Village is located in mountainous areas, so the main material used in construction are the mountain stone.

于家村·清凉阁（井陉县）

又称神仙阁，始建于1581年，此阁上层是砖木结构，下面两层为全石建筑。

Yujia Village · Qing Lang Pavilion (Jingxing County)

Also known as Shenxian Pavilion, built in 1581, the upper of the pavilion built with brick-wood structure, and the lower of the pavilion built with stone.

于家村·观音阁（井陉县）

位于于家村古村落，建于1647年，建筑为单檐，主体为两层。一层为石砌，中有圆拱形门。二层为砖木结构。

Yujia Village · Guanyin Pavilion (Jingxing County)

Guanyin Pavilion locaded in the Yujia village, built in 1647, the main building has a single eave, two layer, the first layer made of stone with a arched door, the secoond layer made of brick-wood structure.

于家村·二层小院（井陉县）

早年间村里所建住宅多为二层，一层用于存放杂物及圈养牲畜，二层是人们居住的屋子，墙体从底到顶都用石材垒筑。

Yujia Village · Two-floor Yard (Jingxing County)

In early days, the village residential construction is mostly two layers, the first layer is used to store the sundries and feed animals, the second layer is used ad the living room, the wall of the yard was built with stone from top to bottom .

于家村·平顶房（井陉县）

村里的建筑很有特色，尤其是住宅。民居多依地势建于石基之上，层层叠叠。多数民居是用石材覆盖混合土建成平顶，可用于晾晒农作物等。

Yujia Village · Flattop (Jingxing County)

The construction of the village is quite distinctive, especially the residences. The residences was built on the base of stone, and the roofs of the houses overlap.The flat topping was built on the composite soil with stones, and people can dry crops on the flat topping.

于家村·大户人家（井陉县）

村里一些大型住宅构造与山西的四合院有相同之处。每户人家均正房、厢房主次分明；部分二层楼房为楼上用砖，楼下用石。

Yujia Village · Rich Families' Residences (Jingxing County)

The structure of some large residences construction had something in common with Siheyuan in Shanxi. Every residence has a main room and wings. A number of the two-story residences have two parts, for the upstairs with brick and the downstairs with stone.

于家村·脚下的窑洞（井陉县）

一些村民在自家门前小路下层挖建类似窑洞似的仓储建筑，用于存放农具、柴草或饲养牲畜。

Yujia Village · The Cave (Jingxing County)

Some villagers dug the Cave which like a storage place under the path in front of their houses, and the cave was used to store agricultural implements , firewood or feed livestock.

于家村·百年石板路（井陉县）

于家村的街、巷、胡同都十分干净，路面是清一色的石头。这些曾经棱角分明的石头，经过村里人几百年的磨踏已十分光亮。

Yujia Village · Centennial Stone Road (Jingxing County)

The streets, lanes and hutongs of the village were all very clean. The surface of the road was covered with stone. These stones had no sharp corners because of hundreds of years.

于家村·农家小院（井陉县）

村里的院落基本都保留了原先的格局，正房、厢房分明，在院落中装点石桌、石凳。由于，此地降水量较少少，因此村里家家都留有旱井。

Yujia Village · Farmyard (Jingxing County)

The Farmyard basically retained the original pattern, and have the main room, the wings, stone tables and stone benches. Because of the less rainfall, every residence had a dry well.

于家村·月亮门（井陉县）

村里有些小阁楼有着江浙建筑的影子，粉墙黛瓦很是醒目。于家村本是于谦后人所建，于谦原籍是浙江，于家村有江浙风格建筑也就顺理成章。

Yujia Village · Moon Gate (Jingxing County)

There are some small attics looks like Jiangzhe's buildings, white wall and black tiles is very smart. Yuqian, the ancestor of the village, built a number of Jiangzhe style building in Yujia Village.

于家村·屋间小路（井陉县）

于家村地处山区，因地势原因，村内民居依高就低顺势而建，交错有秩，而门前小路则随屋而成，蜿蜒于村中，极具特色。

Yujia Village · Path (Jingxing County)

Yujia village is located in mountainous areas, due to the terrain of the village, the dwellings built in accordance with the terrain, and the Path before the dwellings was very characteristics.

当泉村·村口（井陉县）

始建于明朝，村旁山上一股泉水流至村口正好被一块形似卧牛的巨石挡住，村民把这块大石叫做“挡泉石”，当泉村名即是由此而来。

Dangquan Village · The Entrance to the Village (Jingxing County)

Built in the Ming Dynasty. Because of the stone which obstruct the spring water from the village mountain, so the villagers called the stone as"Dang Quanshi", which is the name of the village.

当泉村·观音阁（井陉县）

村里保留着许多古建筑，其中观音阁最具特色，建筑主体为两层，底层为石砌门洞，二层为砖木结构，屋顶覆琉璃瓦。

Dangquan Village · Guanyin Pavilion (Jingxing County)

The village retains many ancient buildings, including the Guanyin Pavilion, which is the most characteristic. The main building of the Guanyin Pavilionis was two layers, the bottom is the stone door, the second layer is brick-wood structure, covered the roof with Liuli tiles.

当泉村 · 老院子（井陉县）

村里古院落一百余处。院落建筑多以石砌为主，辅以青砖，多数院落为典型四合院式。

Dangquan Village · Old Courtyards(Jingxing County)

There are more than one hundred Old Courtyards in the village. And most of the courtyards built with stone and brick, and the majority of the courtyards is a typical Siheyuan type.

当泉村·一街八巷（井陉县）

当泉村一街八巷，所有建筑均依山势而建，呈阶梯状，前排房屋的房顶即可成为后排房屋晾晒农作物或储放柴草的场地。

Dangquan Village · Yijie Baxiang(Jingxing County)

All the streets were built along the mountains shape, showed ladder-like distribution, the front row houses can be used as platforms dry crops and store of straw for the rear houses .

当泉村·山村宅居（井陉县）

远眺依山而建的当泉村，青石砌成的建筑从低到高层层叠叠，极具特色。

Dangquan Village · The Village of Residence (Jingxing County)

Overlooking the Dangquan Village, the bluestone buildings extend from low to high layers, and was very characteristics.

大梁江村·村口（井陉县）

该村形成于明朝，具有六百多年的历史。现有保存较好的明清古民居建筑314座，有162座完整的四合院，是国家级历史文化名村。村口有一小型城门楼。

Daliangjiang Village · The Entrance to the Village (Jingxing County)

The village was formed in the Ming Dynasty, a history of more than six hundred years. There were 314 ancient buildings of Ming and Qing Dynasties which better preservation, and had 162 complete Siheyuan. Daliangjiang Village is a national historical and cultural village.

大梁江村 · 老宅（井陉县）

村里的院落以四合院为主，北面是正房，两侧为厢房，正房中间为客厅，两侧为长辈居住的卧室，南侧的房子背阴，多作为储藏室，宅院大门多开在院子的东南角。

Daliangjiang Village · Old House (Jingxing County)

Most of the village's courtyard was Siheyuan type, which the north was main room and the both sides were the wings, the middle of the main room was the living room, on both wings were the elders' living room. The room on the south side was in the shade, treated as a storage room, and the door of the Old House opened in the southeast corner of the yard.

大梁江村·二层石楼（井陉县）

走在街巷间，大多建筑是二层或者三层的石楼。由于大梁江村古属山西，因此民居具有很多山西建筑的特点，如外墙高大，有很强的防御性等。

Daliangjiang Village · Two-story of Stone Building(Jingxing County)

Walking in the street, most buildings are two or three layers of stone. Because of the Daliangjiang Village belong to Shanxi Province in the ancient time, therefore the residences have many of the characteristics of the construction in Shanxi, such as the tall wall which have very strong defensive function etc..

大梁江村·院间小巷（井陉县）

街道两侧院门各自不同，门窗大多采用石质拱券结构。建筑的形式兼有石头窑、砖木小楼和四合院等多种形式于一体，兼有山西与北京民居的双重特点，同时也有自己的地域特色。

Daliangjiang Village · Alleys (Jingxing County)

Both sides of the alleys have many different gates, a lot of the doors and windows was made of stone and arch structure. The form of the buildings get the form of stone kilns , brick-wood small buildings and Siheyuan into itself, double characteristics of both Shanxi and Beijing dwellings, but also has its own regional characteristics.

大梁江村·武举大院（井陉县）

村里最具特色、保存最完整的是武举大院，大门上悬挂匾额“武魁”。这座大院包含九个小院，俗称“一宅九院”。

Daliangjiang Village · Wuju Courtyard (Jingxing County)

The Wuju Courtyard is the most characteristic and the best preserved courtyard in the village.There is a plaque hanging on the gate carved “Wu kui”. The Wuju Courtyard contains 9 small yards, commonly known as “Yi Zhai Jiu Yuan”.

大梁江村·古村老街（井陉县）

全村建筑依山就次，鳞次栉比，整个村子既没有整齐的街巷，约有3700米村中老街多为宽二米到四米不等的青石小巷，综合交错，别具特色。

Daliangjiang Village · Old Streets of the Village(Jingxing County)

All of the buildings in the village were built down the hillside, row upon row, and the whole village had no orderly street. Old Streets of the Village were about 3700 meters long, ranging from 2 meters to 4 meters wide, formed bluestone lanes, which were very distinctive.

大梁江村·石路石屋（井陉县）

建筑所用石材均选自当地，有的原石未动，有的錾迹斑斑。村庄依山而建，街道高低错落，形成了一个保存完好的古民居建筑群。

Daliangjiang Village · Stone road and Stone house (Jingxing County)

The buildings were built with local stones, and the village was built close to the mountains, streets strewn at random, so forming a well-preserved ancient residential buildings.

大梁江村·屋顶风光（井陉县）

俯瞰古村，建筑高低错落、层层叠叠。以石砌结构、硬山顶建筑居多，同时也有部分平顶建筑。既考虑防火，也考虑农作物的晾晒。

Daliangjiang Village · Roof Scenery (Jingxing County)

Overlooking the village, the majority of the buildings have stone structure and Yingshan roof , and some houses are flat-roofed buildings because of the fire protection and crops drying.

狼窝村·二层小楼（井陉县）

该村形成于明朝，是当地较大的村落。硬山顶建筑居多，建筑的结构以石砌为主。二层小楼的一层常设计成窑洞似，用以储存杂物等。

Langwo Village · Two - storey Building (Jingxing County)

The village was formed in the Ming Dynasty, and was the larger one in the county. Most of the buildings were hard top buildings, and the structure of the buildings was built with stone. The first layer of the Two - storey building often designed like a cave, which was used to store sundries etc..

狼窝村·院门（井陉县）

多数院落呈传统四合院式布局，该院门为拱券式，院门与建筑均为砖石混合结构。下半部为石砌，上半部砖石混合。

Langwo Village · Main Gate (Jingxing County)

The majority of courtyards is the traditional courtyard-style layout, the Main Gate was arch type. The gate and buildings were all brick mixed structure. The lower half part was stone masonry, the upper was brick mixed structure.

河英村·村路（井陉县）

该村的古民居规模均不大，但比较完整，而且精致。村里的房屋材料皆为石材，街道亦全部用青石和卵石铺成，高低错落、纵横交织。

Yinghe Village · Alleys (Jingxing County)

The size of the Ancient dwellings were not big, but relatively complete and exquisite. The houses were stone material, streets were paved with bluestone, scattered high and low.

河英村·门前桥（井陉县）

建筑均依山而建，该建筑院门位置较高，故在门前砌石拱桥一座，穿过桥亦有一座门楼作为出口，别具特色。

Yinghe Village · Bridge (Jingxing County)

The buildings were built close to the mountains, and the position of the building gate was a little higher. So there wasa stone arch bridge in front of the gate , and cross the bridge also has a gate as export, it was very distinctive.

红色漫记

Red Album

中国人民银行旧址（新华区）

又称小灰楼，是一座建于1940年的三层建筑。1948年12月1日中国人民银行总行宣告成立，第一套人民币就曾在这里诞生。现为中国人民银行旧址纪念馆和河北省钱币博物馆。

The Former People's Bank of China(Xinhua District)

Also known as the small gray building, was built in 1940, a three-storey building. In December 1, 1948, the head office of the people's Bank of China was founded, and the first set of RMB was born here. This is former site of the people's Bank of China the museum and the Hebei coin museum.

中央人民广播电台旧址（井陉矿区）

1948年中央人民广播电台由延安迁入井陉矿区，成为当时解放区最大的广播电台，具备了短波发射能力。旧址占地约1.9万平方米，现仅存电台大楼和地下室。

The Central People's Broadcasting Station Site (Jingxing District)

The Central People's broadcasting station moved from Yan'an to Jingxing mining area in 1948, became the largest radio stations in liberated area, with the short-wave transmitting ability. The site covers an area of about 19000 square meters, and it is the only remaining station building and the basement now.

西柏坡·中共七届二中全会旧址（平山县）

西柏坡是著名的革命圣地，中国共产党在这里召开了七届二中全会，会议做出的各项政策规定对新中国的建设有重大作用。

Xibaipo · Plenary Session of the Seventh CPC Site (Pingshan County)

Xibaipo is the famous revolutionary sites, the Communist Party of China held the seven plenary session of the second here and made the policies and regulations have a significant role in the construction of new China.

西柏坡·毛泽东旧居（平山县）

为土坯垒筑的平顶式建筑，建筑面积约120平方米。此住宅分前后两个小院，前院设警卫室，院内有磨盘和猪圈。正房是毛泽东的办公室和卧室。

Xibaipo · Mao Zedong's Former Residence (Xibaipo, Pingshan County)

A flat architecture of adobe rampart, covers an area of 120 square meters. The residence is divided into two small yards, the front set the guard room, the inside has millstone and pigsties. The main room is office and bedroom of Mao zedong.

西柏坡·朱德旧居（平山县）

为三间仿窑洞式建筑，由青石砌成。三个高大的拱形窗户，就像三眼并排的石窑洞，屋内采光好，彼此相通，中间为会客室，两边是办公室和寝室。

Xibaipo · Zhu De's Former Residence (Pingshan County)

As the three imitation of cave architectures, composed of bluestones. There are three tall, arched windows, just like three eyes side-by-side grottoes cave.The house daylighting is good, communicated with each other, the middle room is reception room, on both sides are the office and dormitory.

西柏坡·董必武旧居（平山县）

为四合院式建筑，坐北朝南，建筑面积123平方米，正房为董必武办公和休息的地方。首版人民币上的“中国人民银行”的字样，就是董必武在此居住时提写的。

Xibaipo · Dong Biwu's Former Residence (Pingshan County)

The residence is Siheyuan building, facing south, covering an area of 123 square meters, the main room is Dong Biwu's office and resting place. The words "the People's Bank of China"on the first edition of RMB was written by Dong Biwu in this residence.

西柏坡·任弼时旧居（平山县）

任弼时旧居西邻毛泽东旧居，同为土坯建筑，是北方典型的一个农家四合院形式。正房是他的办公室和寝室。

Xibaipo · Ren Bishi's Former Residence (Pingshan County)

The former residence of Ren Bishi is next to the Mao Zedong's former places of residence, both of the residences are the adobe building, the north typical form of a farmhouse Siheyuan courtyard. The main room is his office and bedroom.

沕沕水发电厂旧址（平山县）

又称沕沕水水电站，建于1947年，是由中国共产党指挥建设的第一座水力发电站。被誉为“红色发电站”、“边区创举”。

Mimi Water Power Plant Site(Pingshan County)

Also called Mimishui hydropower station, built in 1947, is the first hydroelectric power station commanded by the Communist Party of China, also known as the “Red Power stations” or “Border Region initiative”.

民俗漫记

Folk-custom Album

石家庄丝弦（石家庄市）

又名弦腔、河西调等，是在明清俗曲的基础上衍变而来，是河北省特有的古老剧种之一。

Shijiazhuang Si'xian Opera(Shijiazhuang city)

Also known as Xianqiang opera or Hexi tune, is based on the folk song in Ming and Qing Dynasties, and one of the unique ancient drama in Hebei Province.

牛肉罩火烧（石家庄市）

是流传在石家庄地区的一道特色美食。在一锅牛骨汤中泡满火烧，等火烧泡至发涨，捞出火烧切成块放入碗中，同时加入老汤、熟牛肉片、葱花和香菜即成。

Cover Pie Beef (Shijiazhuang city)

It is a special cate spread in Shijiazhuang area. In a pot of beef bone soup, filled with Baked Wheat Cake, remove and cut into pieces into the bowl, and add original soup , cooked beef slices , chopped green onion and parsley at the same time .

藁城宫灯（藁城市）

宫灯是藁城地方特产之一，始于东汉、盛于隋唐，因进贡宫廷故名“宫灯”。

Gaocheng Palace Lanterns(Gaocheng city)

The palace lantern is one of Gaocheng's special local products, began in the Eastern Han, prevailed in the Sui and Tang Dynasties. Because of the lanterns were sent into the palace, so called it "palace lantern".

缸炉烧饼（鹿泉市）

是用火炉烧缸制作的烧饼。用缸制作食品是一种独特的方法，它利用了“缸”内壁的光滑、耐火和厚度，烧出的烧饼不糊、面光，吃起来香、酥、脆，是石家庄特色小吃之一。

Hearth Baked Wheat Cakes (Luquan city)

It used an hearth to baked wheat cakes. ShaoGang is a unique way to prepare food, which used the smooth refractory and the thickness of the lining of the cylinder, burn-out baked wheat cakes just look like the paste surface light, taste sweet and crisp, one of the Shijiazhuang's characteristics snacks.

正定庙会（正定县）

正定的庙会历史悠久，庙会上沧州舞狮、常山战鼓、高跷等非物质文化遗产项目轮番上演，马家卤鸡、八大碗、饸饹、烧卖等传统小吃阵阵飘香。

Zhengding Kermis (Zhengding County)

Zhengding Kermis has a long history, the Intangible Cultural Heritage such as Cangzhou lion, Changshan Zhangu and Gaoqiao show will be on. There is a lot of traditional snacks such as Majia Chicken, Eight Bowl, Heluo and Shaomai and so on.

常山战鼓（正定县）

是中国四大名鼓之首，早在战国时期已具雏形，至明代盛于民间。其表演雄壮威武，气势磅礴，是特色鲜明又具有较高娱乐、审美价值的一种民间艺术项目。

Changshan War Drum (Zhengding County)

It is the first one of China's four famous drums, has been taken shape in the Warring States period and prevailed in the Ming Dynasty. The performence is very mighty and majestic , and has distinctive features , a high entertainment and aesthetic value of a folk art project.

范庄龙牌会（赵县）

范庄人认为自己是勾龙的后代，因此有制成“龙牌”来敬仰供奉的习俗。每年范庄都要举行盛大的祭龙活动，周围各县的民间艺人也来助兴演出，共同表达对龙的崇拜。

Fanzhuang Longpai Kermis (Zhao County)

Fan Zhuang people considered themselves as the descendants of the dragon, therefore they made “dragon brand” to worship their ancestors. Fanzhuang held the kermis in every year, and the surrounding counties folk artists also come to add the grand show, express the worship of the dragon.

栾城抬花杠（栾城县）

为栾城县流传的一种具有地方特色的民间花会舞蹈，源起于人们对苍岩山庙中“三仙姑”的祭祀仪式，以舞者肩抬三仙姑圣像“过驾”而得名，是祭祀舞队中的领队舞种。

Luancheng Taihuagang (Luancheng County)

This is a local characteristics of civil Flower Dance in LuanCheng County, originated in the temple of “SanXianGu” fete ceremony, dancers with shoulder carried SanXianGu’s statue, and it is a leader in the ritual dance team dances.

无极饸饹（无极县）

为河北中部的传统食品，广泛流行于石家庄及周边县市。将和好的荞麦面，放置于一种叫饸饹床的工具中进行压制，压制成型直接煮食，配以特质高汤。

Wuji Heluo (Wuji county)

This is a kind of traditional food in Hebei Province, widely popular in Shijiazhuang and the surrounding cities. Here's how it be made: put uckwheat noodles on Heluo bed, clamped down it and pressure molding directly cooking, with a special stock.

井陉拉花（井陉县）

起源于明清，产生并流传于井陉县境内，是河北最具特色的民间舞种之一。传统井陉拉花有其显著的艺术特色，以“拧肩”、“翻腕”、“扭臂”等动作为主要舞蹈语汇。

Jingxing Lahua (Jingxing county)

Originated in the Ming and Qing Dynasties, produced and spread in Jingxing County, is one of the most characteristic folk dance of Hebei. Jingxing traditional Lahua art has its remarkable artistic features, the "shoulder screw", "palming", "twist", as the main dance vocabulary.

南张井老虎火（井陉县）

“老虎火”是用制作成老虎状的道具放的烟火。南张井村作火技艺距今已有320多年的历史。清朝后期至解放前，南张井村烟火一直是井陉县城元宵花会的压轴戏，现依然表演。

Nanzhangjing Tiger Fire (Jingxing County)

“The tiger fire” is made into a tiger shaped props fireworks. Nanzhangjing Village of fire art has a history of over 320 years. The late Qing Dynasty to the liberation, the tiger fire of Nanzhangjing Village is the last and best item on a theatrical program and is still performing now.

赵庄岭皇纲（井陉县）

又称颠杠，起源于明嘉靖年间，是赵庄岭火帝庙会独具特色的传统节目。因再现古代帝王外巡时的皇家气势而闻名遐迩，倍受人民群众的喜爱。

Zhaozhuangling Huang Gang (Jingxing County)

Also known as Diangang, originated during Jiajing period of Ming Dynasty, a traditional program in Zhaozhuangling kermis. For the reproduction of the emperors when the tour outside momentum Royal and be well-known, far and near, by the people's favorite.

武凡同高跷马（灵寿县）

是一种脚踩高跷、身系竹马，跷、马结合的民间舞蹈。发源并流传于灵寿县武凡同村。一般高跷马有两个角色，扮为男、女二人。

Wufantong Gaoqiaoma (Lingshou County)

It is a foot stilts, which combined the bamboo and stalking-horse together. The folk dance originated and spread in Lingshou County of Wufatong village. General stilts horses have two roles, play as a male and a female.

后　记

从前年春天开始筹备到现在画完本书，我用了将近两年的时间，现在终于可以松口气。看着几本自己认真画完的速写本，亦是一种心理上的小小满足感。全书收录我的钢笔速写一百多幅，还有一些已完成作品因为针对性弱，而未能收入。而很多古迹的各个角度都很漂亮，但因为篇幅问题，只能画一张，亦有很多古迹，在历史上有着重要的地位，但限于表现形式或时间因素而未能进行写生，故留下了一些遗憾，今后有机会定会进行补充。

其实石家庄有着悠久的历史，古建筑资源和民俗传统均极为丰富。在本书的筹备阶段，除一些不利于钢笔速写这种形式进行表现的土层文化遗址等，石家庄的其他省级以上文保单位我均进行了资料的搜集或考察，为此耗费了我大量的精力，现在回想也是一段充实的记忆。钢笔速写是我在工作中经常用的表现形式，但很多情况下都是放松的草稿或意向图的勾勒，也画过很多古迹和民居，但集中于此画自己家乡的传统风貌，还是第一次。在绘画的过程中，为客观真实的呈现本土建筑及民俗风貌，我尽量让自己的钢笔线条工整，但要避免呆板，又要让画面和线条呈现出个人风格和色彩。为此，我尝试了很多钢笔及纸张，同时，为了让不同的古迹类别呈现不同的风格，建筑类与民俗类呈现不同风格，也费了很多心思。现在看来，效果是有一些的，但也有些地方方面感觉画的不够好，希望今后能够改正。

一直以来，我都认为钢笔速写不应仅仅是画画，更是通过速写的写生内容，认识一种文化，我希望通过每一幅画都能对这些古建筑有一个文化上的推广，画者不应仅仅是画者，更应是文化的传播者。

本书的筹备和写生过程时间紧迫，亦有不足之初，还恳请读者和同行指正，在此深表谢意。

最后需要说明的是，在本书的筹备和书写过程中，我的家人给予了我巨大的支持和帮助，借此机会向他们表示深深的谢意！

宋 端

2013年5月